I0805187

Nature's Mysteries

NAMIBIA'S FAIRY CIRCLES

by Matt Lilley

abdobooks.com

Published by Pop!, a division of ABDO, PO Box 398166, Minneapolis, Minnesota 55439. Copyright © 2021 by POP, LLC. International copyrights reserved in all countries. No part of this book may be reproduced in any form without written permission from the publisher. Pop!™ is a trademark and logo of POP, LLC.

Printed in the United States of America, North Mankato, Minnesota.

102020
012021

THIS BOOK CONTAINS RECYCLED MATERIALS

Cover Photo: iStockphoto

Interior Photos: iStockphoto, 1, 5, 6, 9, 11, 18, 27; Shutterstock Images, 7, 8, 14, 22–23 (grass), 22–23 (termites), 22–23 (dirt), 22–23 (roots), 22–23 (water), 26, 29; Maggie Janik/age fotostock/SuperStock, 12–13; robertharding/Alamy, 15; NHPA/Photoshot/Science Source, 17; Nigel J. Dennis/Science Source, 19; Danit Kymri Wilt/Delimont Creative/Alamy, 21; Edwin Remsberg/VWPics/AP Images, 25

Editor: Alyssa Krekelberg

Series Designers: Candice Keimig, Victoria Bates, and Laura Graphenteen

Library of Congress Control Number: 2020940292

Publisher's Cataloging-in-Publication Data

Names: Lilley, Matt, author.

Title: Namibia's fairy circles / by Matt Lilley

Description: Minneapolis, Minnesota : POP!, 2021 | Series: Nature's mysteries | Includes online resources and index

Identifiers: ISBN 9781532169205 (lib. bdg.) | ISBN 9781532169564 (ebook)

Subjects: LCSH: Geology--Namibia--Juvenile literature. | Termites--Juvenile literature. | Desert ecology--Juvenile literature. | Geography--Juvenile literature. | Curiosities and wonders--Juvenile literature. | Mystery--Juvenile literature.

Classification: DDC 910.02--dc23

WELCOME TO DiscoverRoo!

Pop open this book and you'll find QR codes loaded with information, so you can learn even more!

Scan this code* and others like it while you read, or visit the website below to make this book pop!

popbooksonline.com/fairy-circles

*Scanning QR codes requires a web-enabled smart device with a QR code reader app and a camera.

TABLE OF CONTENTS

CHAPTER 1

A MYSTERY IN THE DESERT

A scientist walks across a stretch of sand. She looks down. On the ground are strange patterns. Tall grasses grow in circles. Inside the circles, there is nothing but sand.

WATCH A VIDEO HERE!

Scientists have been studying fairy circles since the early 1970s.

Some people say Namibia's fairy circles are one of nature's greatest mysteries.

The strange patterns are called fairy circles. They appear throughout a strip of land in the Namib Desert. They cover approximately 1,118 miles (1,800 km). The desert is located in southwestern Africa. It runs along the Atlantic Ocean.

FUN FACT

In the Nama language, *Namib* means "vast place."

WHERE IS THE NAMIB DESERT?

Not many people live in the Namib Desert. The climate is too harsh.

No one is sure why the plants grow like this. But scientists have a few ideas. For instance, the fairy circles might be related to the area's **climate**. The Namib Desert is very dry. Its plants and animals have **adapted** to live without much water.

A gemsbok can go several days without taking a drink. It gets most of the water it needs by eating plants or licking up dew.

CHAPTER 2

ARE PLANTS CAUSING IT?

Water in the desert is hard to find. The Namib Desert gets less than 4 inches (10 cm) of rain each year. When there's not enough water, plants **compete** for it. Some scientists think this competition creates fairy circles.

LEARN MORE HERE!

Plants will die if they don't get enough water.

Grasses make shade and keep the ground around them moist. This helps more grasses grow nearby. They form clumps. The grasses in clumps grow better than the grasses elsewhere.

Scientists studied why grasses seemed to grow everywhere but inside the circles.

FUN FACT

Fairy circles don't last forever. Eventually, grasses will fill in the bare patches. Then the pattern disappears. Most circles last 40 to 60 years.

Plants use their roots to suck up nutrients and water from the soil.

The grasses clumped together have longer roots. So they get more water than other grasses. In dry weather, the grasses with short roots outside the clump die. Over time, several clumps of surviving grasses could make a circle.

Some people say the circles look like they've been created by meteorites crashing into Earth.

CHAPTER 3

ARE TERMITES CAUSING IT?

Some scientists think termites make the fairy circles. Termites eat grasses. Sometimes they eat all the grasses in a circular area around their nests. This can make a grassless circle.

COMPLETE AN ACTIVITY HERE!

Some scientists believe harvester termites create the circles. But others wonder whether different termites are responsible.

Deserts don't get much rain. But when rain does fall, it tends to be heavy.

Getting rid of the grasses lets more rainwater sink into the soil. With no grasses inside the circle to drink the water, more water stays in the ground. The termites let the grasses outside the circle reach the water. These healthy grasses can be food when the termites need it. By making the grasses grow in these patterns, the termites can store food and water.

Harvester termites dig their nests underground. They visit the surface to find food.

AUSTRALIA'S FAIRY CIRCLES

In 2016, scientists announced that they had discovered fairy circles in Australia. These scientists looked for termites but didn't find many. They said this meant that the plants caused the circles. Other scientists disagreed. They thought termites caused these fairy circles too. The **debate** continued.

If termites make the fairy circles, then all fairy circles must have termites. But during research, scientists couldn't find termites in every circle. Some scientists say they just need to look harder.

Not all scientists agree on why the fairy circles appear in strange patterns.

FUN FACT

Fairy circles appear in roughly **hexagonal** patterns. One circle will usually be surrounded by six other circles.

TERMITES AND FAIRY CIRCLES

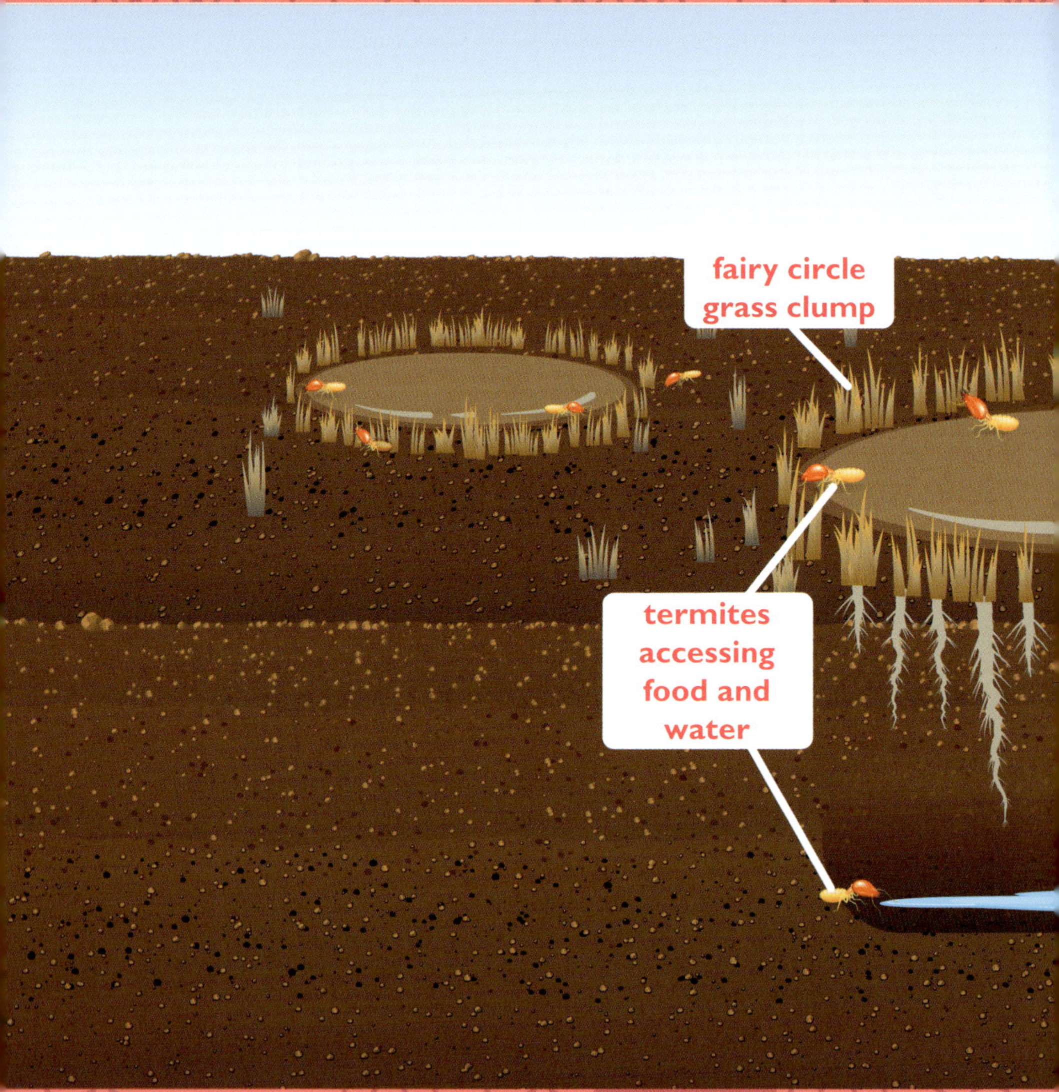

Some scientists think termites won't let plants grow in the middle of a fairy circle. This allows rainwater to sink into the soil there. Termites then drink the water. They also let the fairy circle grass clumps access the water. That way, the grass clumps live, and the termites can eat the grasses.

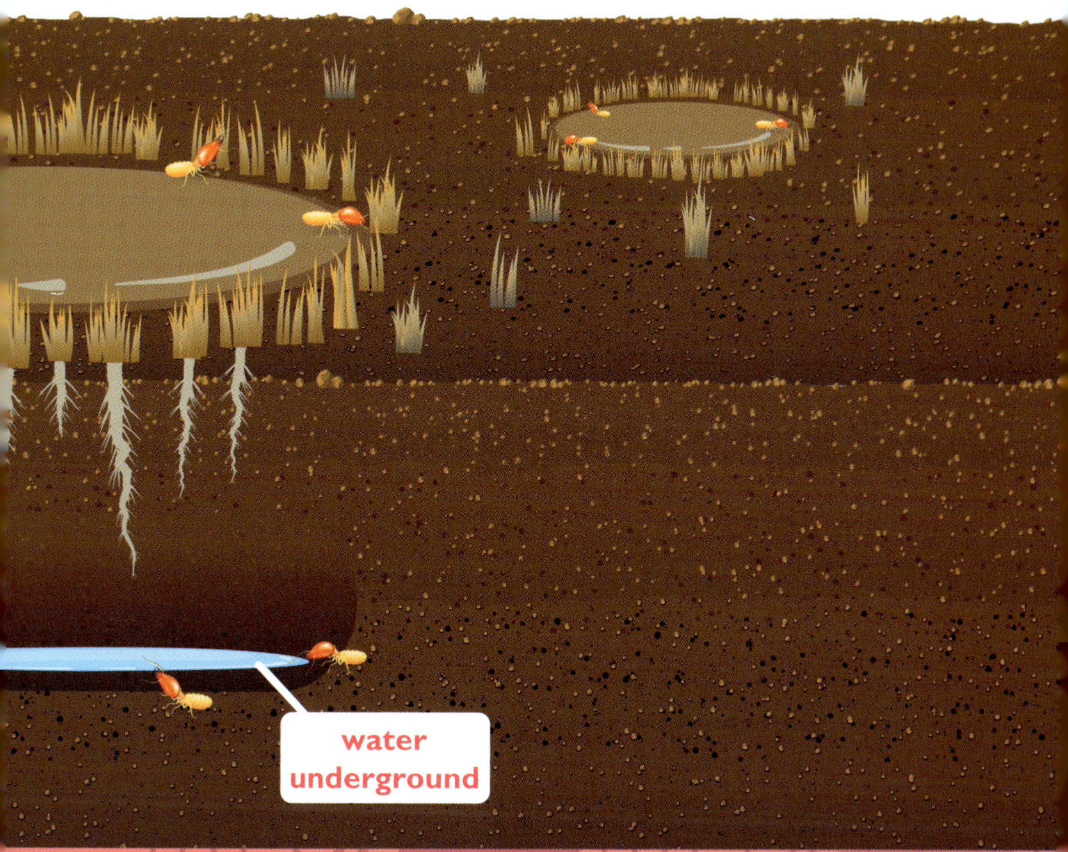

CHAPTER 4

A COMBINED CAUSE

Some scientists think termites and plants might create the circles together. Scientists studied the circles. They also studied the patterns around the circles. They used computers to **analyze** the patterns.

LEARN MORE HERE!

People travel deep into the Namib Desert to study the mysterious circles.

Scientists from around the world have studied fairy circles.

The computer analysis supported this idea. Plants on their own could only cause some of the patterns. The same was true with termites. But together, termites and plants could cause all of the patterns.

Plants and animals often work together. For example, bees get food from flowers and spread the plants' pollen.

However, the mystery isn't completely solved. Scientists are still **debating**. Some think only the grasses are making the patterns. Some say it's only the termites. Some agree it's both, but they argue over how much is caused by grasses or by termites. More study is needed to fully solve this mystery.

FUN FACT

The Himba people who live in this area have a legend about how the circles were made. They say the circles are the footprints of a god.

Some people claim the circles were made by a dragon breathing fire.

MAKING CONNECTIONS

TEXT-TO-SELF

Which explanation of how fairy circles form do you think is most likely to be correct?

TEXT-TO-TEXT

The fairy circles appear in the Namib Desert. Have you read other books about deserts? What did you learn?

TEXT-TO-WORLD

Computers help scientists analyze the patterns of fairy circles. Can you think of other tasks scientists use computers for?

GLOSSARY

adapt — to develop traits that make it easier to survive in an environment.

analyze — to study carefully.

climate — the typical weather of a place or area over time.

compete — to try to get something that others are trying to get too.

debate — to argue for or against something.

hexagon — a six-sided shape.

INDEX

ONLINE RESOURCES

popbooksonline.com

Scan this code* and others like it while you read, or visit the website below to make this book pop!

popbooksonline.com/fairy-circles

*Scanning QR codes requires a web-enabled smart device with a QR code reader app and a camera.